Water
221

从海洋到天空

From Sea to Sky

Gunter Pauli

［比］冈特·鲍利 著

［哥伦］凯瑟琳娜·巴赫 绘

李原原 译

上海远东出版社

丛书编委会

主　任：贾　峰

副主任：何家振　闫世东　郑立明

委　员：李原原　祝真旭　牛玲娟　梁雅丽　任泽林

　　　　王　岢　陈　卫　郑循如　吴建民　彭　勇

　　　　王梦雨　戴　虹　靳增江　孟　蝶　崔晓晓

特别感谢以下热心人士对童书工作的支持：

匡志强　方　芳　宋小华　解　东　厉　云　李　婧

刘　丹　熊彩虹　罗淑怡　旷　婉　杨　荣　刘学振

何圣霖　王必斗　潘林平　熊志强　廖清州　谭燕宁

王　征　白　纯　张林霞　寿颖慧　罗　佳　傅　俊

胡海朋　白永喆　韦小宏　李　杰　欧　亮

目录

Contents

ZURI Learning Initiative

一条肺鱼看到几只鸡正试图往树上爬，准备在树枝上安稳地过个夜。

　　"你知道你是我的后代吗？"肺鱼对其中一只鸡大声喊道。

A lungfish is watching some chickens trying to get up into a tree, to pass the night there, safe on its branches.

"You know you are a descendent of mine?" Lungfish calls out loudly to one of the chickens.

一条肺鱼看到几只鸡……

A lungfish is watching some chickens ...

不会受到狐狸的伤害……

Safe from any foxes …

"抱歉，先让我在树上把宝宝安顿好，这样我们就不会受到狐狸的伤害了。"鸡回应道。

"谨慎点总没错。你知道吗？如果不是我喜欢在水中保持平衡的话，你们就不可能在陆地上安居。"

"Excuse me, let me just get my chicks up into the canopy where we will be safe from any foxes," the chicken responds.
"Better to be safe than sorry. Do you know that you would never have succeeded in getting to safety on land, if I had not been so keen on keeping my balance in the water?"

"我确实欣赏你和你所做的事，但声称我是你的后代？然后我应该感谢你有肺？这种说法至少是有些夸大了。"

"我知道，当一个人听到一个从未听说过的消息时，他的第一反应是：不是真的！不可能的！"

"I do appreciate you for who you are, and for what you do, but to claim that I am a descendent of yours? And then that I should thank you for having lungs? That is, to say the least, exaggerated."

"I know that whenever presented with a piece of information one has never heard before, one's first reaction is: Not true! Not possible!"

我是你的后代？

I am a descendent of yours?

我的鱼鳔变成了你的肺……

My swim bladder turned into your lungs ...

"那么，你是只想追求名声呢，还是愿意花时间解释一下呢？"

"我的鱼鳔变成了你的肺。"

"那又怎样？那也不会让我成为你的后代呀。"鸡回应道。

"So, will you only claim your name and fame, or will you take the time to explain?"

"My swim bladder turned into your lungs."

"So what? That does not make me your offspring," Chicken replies.

"如果你想离开水到陆地上生活，你需要一个能够从空气中获取氧气的器官，而不是通过过滤水中的微小气泡来获取氧气。"

　　"当然。但你怎么解释我的脖子？你的脖子固定在肩上不能动，而我的脖子可以让我上下左右张望。"

"For you to get out of the water and live on land, you needed a device that could take oxygen from the air, instead of filtering tiny bubbles of it from the water."

"Granted. But then how do you explain my neck? Your neck is fixed to your shoulders while my neck lets me look up and down, left and right."

你怎么解释我的脖子？

how do you explain my neck?

我不需要靠骨头把身体各部位连接起来……

Didn't need bones to hold my body together …

"你看，我在水里游泳，在不受重力束缚的情况下可以四处移动，我不需要像你们鸡一样，靠骨头和韧带把身体各部位连接起来。"

"你知道我们离开水面，开始在陆地上行走后，我的祖先是怎么开始在空中飞行的吗？"

"Look, as I was swimming in the water, and moving around without gravity, I didn't need any of the bones and ligaments you chickens have to hold my body together."

"And do you know how my ancestors got to fly in the air, after we got out of the water and started walking on land?"

"我知道，一开始，你的每个下肢都有8个脚趾，由于大多数都没用，最后只留下四五个。但你们是如何飞到空中的，这对我来说是个谜。"

"那两条腿是用来跑的，不是用来飞的。"

"那么，你是怎么知道你可以获得升力来克服重力，进而飞起来的呢？"

"I know you, at first, had eight toes on each lower limb, and since most were useless, you ended up with only four or five. But how you became able to lift off into the air is a mystery to me."

"Those two limbs were used for running, not for flying."

"So how did you figure out you could get the lift to defy gravity and fly?"

是用来跑的，不是用来飞的……

Used for running, not for flying…

扇动我们的短翅膀......

flapping our short wings ...

"当我们爬到树上时，我们会很容易被重力拉下来。但通过疯狂地扇动我们的短翅膀，我们就'粘'在树上了——就像赛车的扰流板把疾驰的车辆压到赛道上一样。"

"啊，所以你的飞行是从爬树开始的，那可是一种在夜晚决定生死的做法。"

"When we ran up a tree, we would be easily pulled down again by gravity. But by flapping our short wings like crazy, we 'stick' to the tree – just like the spoiler of a racing car presses the speeding vehicle to the track."

"Ah, so your flying started when you made your vertical run up trees, something that made the difference between life and death at night."

"接下来，我们长出羽毛来保暖。当然，一旦我们有了羽毛，我们就开始用它们的色彩来吸引女士们。"

"然后你就开始追逐飞虫，张开翅膀追逐它们。在你快速追赶的过程中，你跳到了空中……"

"我们的羽毛也给了我们额外的升力！"

"在你们意识到这一点之前，你们就已经是能克服重力的飞鸟了。这也需要用到我给你们的肺！"

……这仅仅是开始！……

"Next, we developed feathers to keep us warm. And, of course, once we had them we started using them to impress the ladies with our colours."

"Then you started running around after flying bugs, pursuing them with outstretched wings. During your speedy chase, you leaped into the air…"

"And our feathers gave us that extra lift!"

"Before you knew it, you were birds defying gravity. With the lungs you got from me!"

... AND IT HAS ONLY JUST BEGUN!...

……这仅仅是开始！……

…AND IT HAS ONLY JUST BEGUN! …

Did You Know ?

你知道吗?

肺

Land vertebrates use their lungs for breathing. A few groups of fish like lungfish, which were the likely ancestors of land vertebrates, also have lungs.

陆地脊椎动物用肺呼吸。一些鱼类，比如肺鱼，可能是陆地脊椎动物的祖先，也有肺。

Darwin believed that lungs evolved from gas bladders. However, fish with lungs are the oldest type of bony fish. The lung comes from the tissue sac that surrounds the gills.

达尔文认为肺是由鱼鳔进化而来的。然而，有肺的鱼是最古老的硬骨鱼。肺来自鳃周围的组织囊。

在 3.9 亿年前至 3.6 亿年前，肺鱼的后代开始在较浅的水域生活，然后迁移到陆地。这些鱼长着肉鳍，后来变成了人类身上的四肢。

Between 390 and 360 million years ago, the descendants of lungfish began to live in shallower waters, and then moved onto land. These fish had lobed-fins, which turned into limbs in humans.

在鱼类变成陆生生物之前，辐鳍鱼的祖先就有肺和鳃。鱼的肺后来进化成鱼鳔——一个充满气体的器官，帮助鱼控制自己的浮力。

The ancestor of the ray-finned fish had lungs and gills before fish turned terrestrial. Lungs in fish later evolved into the swim bladder, a gas-filled organ that helps the fish control its buoyancy.

The impulse to breathe, sleep, and move, is given to us at birth. It takes practice, as well as the motivation from its parents, to help a baby to learn to walk, and reach its full potential.

我们出生时就有呼吸、睡眠和运动的冲动。帮助婴儿学会走路并发挥其全部潜能既需要练习，也需要父母的鼓励。

Early birds, which are descendants of dinosaurs, may have used their wings not for flying, but for running. By flapping their front appendages, the animals gained more traction as they were running up steep inclines.

早期的鸟类是恐龙的后代，它们的翅膀可能不是用来飞行的，而是用来奔跑的。通过拍打前肢，它们在陡峭的斜坡上奔跑时能获得更大的牵引力。

大多数鸡有 4 个脚趾。它们的肌肉系统就像一把锁，使它们能够在栖木上睡觉。但也有例外：丝羽乌骨鸡、杜金鸡、法夫罗鸡和胡登鸡都有 5 个脚趾。

Most chickens have four toes. Their muscular system works like a lock to enable it to sleep on a perch. There are exceptions: Silkies, Dorking, Faverolle, and Houden chickens have five toes.

The chicken has its origin in a group of dinosaurs called the theropods. Genetically speaking, the domestic chicken is the closest living relative of Tyrannosaurus rex.

鸡起源于兽脚亚目恐龙。从基因上来说，家鸡是霸王龙现存的最近的亲戚。

Think about It

想一想

Does it sound reasonable that a chicken is family of the T. Rex?

鸡是霸王龙的亲戚，这听起来合理吗？

And what about a fish being the forefather of a chicken?

鱼是鸡的祖先，这听起来合理吗？

When you hear something new for the first time, do you think it is impossible?

当你第一次听到一个新鲜事时，你会认为这是不可能的吗？

Ever thought about running up a tree?

想过爬上树吗？

Do It Yourself!

Find out if the people around you know about evolution. Charles Darwin may be a name most people will recognise. Prepare a short and simple explanation of how a chicken is related to a fish and a dinosaur, the T. rex. In first explaining this to others, you will, most likely, be met with disbelief. So prepare a further explanation that will make it more credible, so you will receive less and less resistance to this concept. When you think you have your arguments well lined-up, test them on friends and family members, until they are as convinced as you are.

看看你周围的人是否知道进化论。查尔斯·达尔文可能是大多数人都知道的名字。准备一份简短的资料，说明鸡、鱼、霸王龙是如何联系在一起的。在第一次向别人解释这个问题时，对方很可能会感到难以置信。所以准备一个更详尽、更可信的说明，这样别人就会渐渐地不再那么排斥这一想法。当你认为论据已经很充分时，试着说服朋友和家人，直到他们和你一样信服。

27

学科知识
Academic Knowledge

生物学	鸡和肺鱼是杂食动物；肺鱼有肺和鳃；干旱时，肺鱼会用鱼鳔呼吸，当水干涸时，肺鱼就会冬眠或在干燥坚硬的泥土中保持休眠状态；鸡有自己独特的语言，会用超过30种不同的声音来交流；脚的进化。
化 学	鸡能尝出咸味，但尝不出甜味；在美国销售的抗生素有80%用于牲畜。
物 理	浮力和表面张力的影响；当气球排开和自身质量相等的空气后，它会停止上升；奔跑时利用前附肢使身体始终贴近地面。
工程学	用压载物控制浮力；即使是钢（密度比水大得多）制的船，也能浮在水面上，因为钢制的船封闭了大量的空气（密度比水小得多），而且船的平均密度比水小；扰流板把赛车压到赛道上。
经济学	通过选择性育种，养鸡场的鸡长得又大又快，以原先一半的时间达到出栏体重。
伦理学	随着耐抗生素疾病的出现，消费者一直在呼吁将这些药物撤出农场；在鸡笼中养鸡。
历 史	肺鱼在3.9亿年前到3.6亿年前的泥盆纪时期从海洋迁移到陆地。
地 理	肺鱼生活在非洲、南美洲和澳大利亚。
数 学	阿基米德原理：浮力等于物体排开的流体的重力。
生活方式	地球上的鸡比人还多（250亿只鸡）；表明感激之情；知道自己卑微的出身；人们不愿花时间去解释，只会用口号来表达。
社会学	许多人喜爱鸡，因为它会在黎明前啼叫，预示着黎明的到来。
心理学	养鸡场的鸡生活条件非常恶劣，以致同类相食；面对坏消息时怀疑的反应。
系统论	鸡能帮我们消灭虫子，还能为我们提供鸡蛋；大自然总是有多种功能，例如羽毛既能保暖，又能提供升力。

情感智慧
Emotional Intelligence

肺　鱼

肺鱼想在一些事上说服鸡，就大声招呼，来引起她的注意。他不断强调自己的论点，并傲慢地回应。当鸡礼貌地回答并提出质疑时，肺鱼给了一句非常简短的回应，希望能给鸡留下深刻的印象。当鸡和他互换角色，开始谈论自己时，肺鱼向她展示自己是多么见多识广。然后，他积极互动，一步一步紧跟鸡的逻辑，展示他对论点的警觉性和清晰的思维。肺鱼证明了他对自己的角色有自知之明，并很自豪地与他人分享。

鸡

鸡礼貌地要求肺鱼先给她时间照顾自己的小鸡。她要求肺鱼耐心，并带着同理心说话。她清楚地陈述了自己的需求，并有信心直接分享自己的观点。她展示了自己的口才。她希望肺鱼能够证实他的论点。在收到回复后，她试图挑战肺鱼，坚持要求肺鱼提供更多的论据。她用令人困惑的信息来否认肺鱼。不过她那引人入胜的陈述事实的方式，确实在他们最初不舒服的交流之后，使得肺鱼加入积极的对话并产生同理心。

艺术
The Arts

让我们演示一下"进化"。这就是"变形金刚"的工作。我们都见过这样的图片：一只用四肢行走的猿猴进化成一个会走路的原始人，然后变成一个现代的人。现在让我们想象一下从鱼变成鸟的六个步骤。想象一下这个过程是如何发展的，然后画一些简单的草图来说明你的想象。绘画时要重点关注进化的动力。

思维拓展
Systems: Making the Connections

对进化的研究已经发展成为一门独立的科学。我们可以越来越多地追溯我们从哪里来。除了了解每种生物的起源和历史，我们还能了解它们存在的原因。大自然的进化去掉了多余的东西。任何无法适应环境的生命形式都不会存在太久。然而，大自然也改变了现有的东西，提供了新的功能，这些功能是以前想象不到的。控制浮力的鱼鳔变成了肺，前肢变成了翅膀。这种进化的能力是生物进化的关键特征之一——总是适应新的现实，总是准备好把任何状况变成一个新的优势。恐龙不是生来就会飞的，但是鸡却利用已有的四肢变成了会飞的鸟。当我们能更好地理解我们是如何进化的，我们就会对我们最终的归宿有一些领悟。我们有着这样一种文化和传统，即渴望了解自己从哪里来，但我们也要了解我们该往哪里去。我们需要获得更强的适应能力以适应气候变化，从而减少不确定性对我们的影响。通过进化，鸟类在树上避难的能力增强了，即使它们睡着了，它们的脚爪也能把它们固定在晚上栖息的树枝上。然而，肺鱼作为最早拥有肺的物种之一，长久以来没有任何重大的变化，但仍然保持了自己在生态系统中的位置，并蓬勃发展。既要加快适应，又要保持稳定，没有一条规则可以指导所有的进化过程。很明显，我们现在比以往任何时候都更有雄心，更有想象力和创新精神。从鸡的角度来看，它能在离开浅滩之前，想象自己会变成一只鸟，并且在那个没有鸟、没有翅膀、没有羽毛的时代学会飞翔吗？虽然在事后想象和解释这些进化过程很容易，但要事先进行想象，并把想象变成现实，就比较困难了。

动手能力
Capacity to Implement

我们使用的每一样东西，每一件艺术品，都是在某个时间点被想象出来的。现在，我们需要专注于今后的想象和创造。让我们以飞机为例。你能想象一架新型的飞机吗？删除所有不需要的部件，并决定应该添加什么。让我们设想一种新的交通方式。要有信心，要敢于超越"标准"。现在给你的项目设定一个时间表：它什么时候能实现？

故事灵感来自
This Fable Is Inspired by

美琳娜·E·黑尔
Melina E. Hale

美琳娜·E·黑尔1992年毕业于美国北卡罗来纳州达勒姆的杜克大学，获得动物学学士学位。1998年，她在芝加哥大学获得生物力学博士学位。在马萨诸塞州伍兹霍尔海洋生物实验室格拉斯实验室完成一项神经生物学研究后，她于2001年在纽约州立大学从事神经生物学方面的博士后研究。她现在是格罗斯曼神经科学研究所的教授，并在芝加哥大学教授有机生物学，在那里她主管黑尔实验室。

图书在版编目（CIP）数据

冈特生态童书.第七辑：全36册：汉英对照 /
（比）冈特·鲍利著；（哥伦）凯瑟琳娜·巴赫绘；
何家振等译.—上海：上海远东出版社，2020
ISBN 978-7-5476-1671-0

Ⅰ.①冈… Ⅱ.①冈… ②凯… ③何… Ⅲ.①生态
环境–环境保护–儿童读物—汉英 Ⅳ.①X171.1-49

中国版本图书馆CIP数据核字（2020）第236911号

策　　划　张　蓉
责任编辑　祁东城
封面设计　魏　来　李　廉

冈特生态童书

从海洋到天空

[比]冈特·鲍利　著
[哥伦]凯瑟琳娜·巴赫　绘

李原原　译

记得要和身边的小朋友分享环保知识哦！
八喜冰淇淋祝你成为环保小使者！